Picture Puzzles

with

Cuisenaire® Rods

by
Patricia S. Davidson
Jeffrey B. Sellon

Cuisenaire Company of America, Inc.
12 Church Street
New Rochelle, N.Y. 10805

TABLE OF CONTENTS

INTRODUCTORY NOTES

Picture Puzzles with Cuisenaire Rods is a different kind of crossword puzzle book. It combines two familiar uses of Cuisenaire rods for an exciting, self-checking, computational skills workbook. First, the rods are used as a model to help children solve problems in basic arithmetic. Then the rods are used to record answers to the problems and to create pictures and designs.

The steps necessary to complete each of the worksheets are basically the same for each rod picture:

4 Remove the rods from the ROD WORK AREA and follow the same procedure to answer the remaining problems under ACROSS.

5 Do each problem from the DOWN column, recording the numerical answers and the rod answers. Be sure to place the rod answers DOWN (vertically), on the gridded portion of the sheet.

3 Place the single rod which represents the answer to **a** ACROSS (horizontally) on the gridded portion of the sheet.

6 Check to see if your completed picture looks right. If any rod seems misplaced, go back and check your work.

2 Using the Rule: White = 1, record the numerical answer.

1 Solve problem **a** by placing the correct rods in the ROD WORK AREA.

7 Children may wish to color the rod picture. Suggest they remove one rod at a time and color the correct number of centimeter squares using matching color crayons.

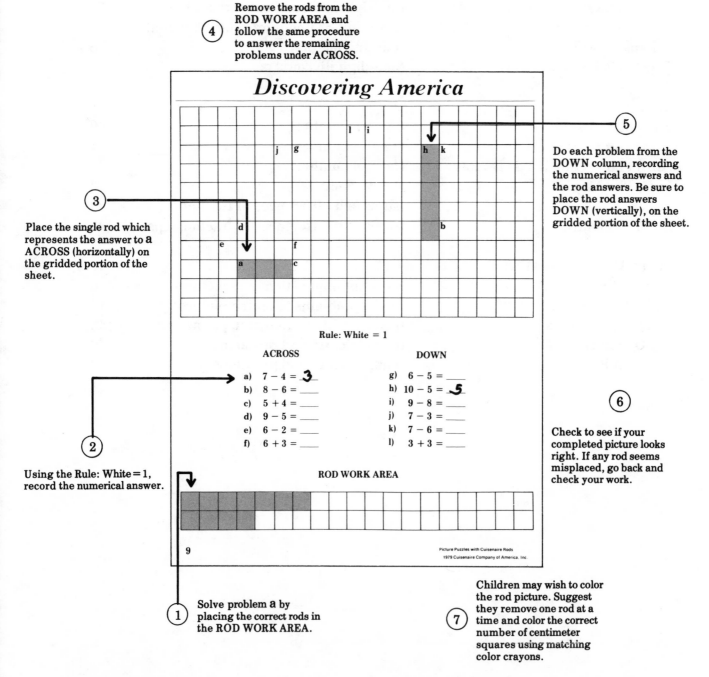

Discovering America

Rule: White = 1

ACROSS

a) $7 - 4 = $ **3**
b) $8 - 6 = $ ____
c) $5 + 4 = $ ____
d) $9 - 5 = $ ____
e) $6 - 2 = $ ____
f) $6 + 3 = $ ____

DOWN

g) $6 - 5 = $ ____
h) $10 - 5 = $ **5**
i) $9 - 8 = $ ____
j) $7 - 3 = $ ____
k) $7 - 6 = $ ____
l) $3 + 3 = $ ____

ROD WORK AREA

9

Picture Puzzles with Cuisenaire Rods
1979 Cuisenaire Company of America, Inc.

NOTE: The next pages describe how to use the rods in the ROD WORK AREA to solve each type of mathematical problem.

HOW TO USE CUISENAIRE® RODS

Cuisenaire® rods come in ten lengths and ten colors. Number values for each are determined by measuring one rod by another. Most often, children measure all of the rods in the set with white rods. When this is done, the other rods assume the values 2 through 10. Picture Puzzles with Cuisenaire Rods uses only white rods to represent 1. In other contexts, children may wish to measure the other rods, for example, with orange. When this is done, the white rod is seen as 1/10 of the orange rod, and the other rods represent the fractions 2/10 through 9/10.

The rod staircase is helpful as a reference in knowing the color names and sequence.

Rod Staircase	Color Code Name	Numerical Value When White = 1
	White	1
	Red	2
	Green	3
	Purple	4
	Yellow	5
	Dark green	6
	blacK	7
	browN	8
	bluE	9
	Orange	10

The underlined letters are the standard Rod Codes.

Cuisenaire rods help to make the arithmetic operations concrete and visual. To use rods successfully, there are only a few basic conventions that need to be learned. Rods are placed end-to-end in a "train" to model addition. If the "cars" of the train are all the same color, then they can be seen as repeated addition, or multiplication. Rods are placed side-by-side for comparisons, subtraction, and division problems. One problem of each kind is shown below, with specific references to pertinent pages in this book.

Addition (pages 6-11, 24-27)

Suppose the addition exercise is 6+2. Since white=1, 6 white rods or 1 dark green rod can be used to represent 6; 2 white rods or 1 red rod can be used for 2.

Addition is solved by placing rods end-to-end on the ROD WORK AREA at the bottom of the picture puzzle page.

Step 1: Place a dark green rod and a red rod end-to-end.

6+2

Step 2: Find the single rod that matches the sum. The answer is a brown rod.

6+2=8

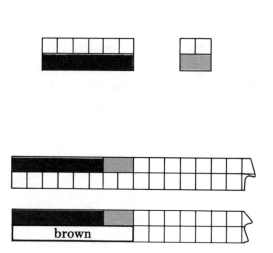

Subtraction (pages 6-11, 24-26)

Suppose the subtraction exercise is $7-4=$ _____.
The first step in a subtraction problem is to put the appropriate rods side-by-side.

Step 1: Place a black rod (representing 7 whites) and a purple rod (representing 4 whites) side-by-side.
$$7-4$$

Step 2: Find the single rod that shows the difference. The answer is a green rod.
$$7-4=3$$

Inequalities (pages 12-16)

The rod staircase is probably the most helpful way to see what rods are one more than or one less than any given rod. The other way of doing these problems is to add (for one more) or subtract (for one less) a white rod. Also the staircase shows what rod is between two rods. For example, the blue is between brown and orange.

Placing two rods side-by-side shows clearly which one is longer (or shorter). This inequality relationship can be translated into numbers.

A yellow rod is longer than a purple rod. With white = 1, this shows 5 > 4, read "5 is greater than 4."

A purple rod is shorter than a yellow rod. With white = 1, this shows 4 < 5, read "4 is less than 5."

Place Value (pages 17-19)

Numbers beyond 10 can be expressed as "Orange Plus" trains. For example,
$$14=10+4$$

Missing Addends (pages 20-23)

The problem, $5+$ _____ $=7$, is read "5 plus what number equals 7." Fortunately with rods, the 5 and 7 both can be shown, and the "missing addend" is visually clear.

Step 1: Yellow + _____ = black.

Step 2: Find the missing rod. The missing rod is red. The numerical answer is 2.

Two-digit Addition (pages 28-33)

A different type of ROD WORK AREA is used for the two-digit addition and subtraction work. The Place Value Mat has a place for tens (orange rods) and ones.

Tens	Ones
orange	black
orange	
orange	dark green

27
+16

Step 1: Place the appropriate rods on the mat to represent the two addends.
Step 2: Add the ones and trade for an "orange plus" train if possible.

black	dark green
orange	green

Step 3: After the trade, there should be 4 orange rods and 1 green rod on the Place Value Mat. The numerical answer is 43.

Two-digit Subtraction (pages 30-33)

For two digit subtraction work, place only one collection of rods on the mat. From this collection, subtract the other collection. Regrouping may or may not be required. For example,

$$35 - 12$$

Step 1: Place the needed rods to represent 35 on the mat.

Tens	Ones
orange	yellow
orange	
orange	

Step 2: To subtract 12, subtract the "ones" by subtracting yellow – red. The answer is a green rod, or 3 "ones."

$$\begin{array}{r} 35 \\ -12 \\ \hline 3 \end{array}$$

Step 3: Now subtract 1 "ten" from the 3 "tens." The answer is 2 orange rods.

$$\begin{array}{r} 35 \\ -12 \\ \hline 23 \end{array}$$

Multiplication (pages 34-39)

The simplest rod model for one-digit multiplication is the "repeated addition" train. For example, 3×4 can be shown as 3 purple rods placed end-to-end.

Since the 3 purple rods can be matched with orange + red, the numerical answer in terms of white rods is 12.

Division (pages 34-39)

Division problems are solved in a similar way. $12 \div 4$ can be interpreted as "how many 4's are there in 12?"

Step 1: Start with an orange + red train.

Step 2: Match the train with purple rods. How many does it take? Is there a remainder?

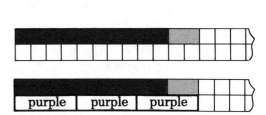

Fractional Parts (pages 40-45)

There is another way to interpret division. Since it takes 3 purple rods to match 12 whites, each purple rod is ⅓ of 12.

In doing problems like $(½ \times 12) + (⅓ \times 12)$, children should do each fraction problem separately and then add the results:

Step 1: $½ \times 12 = 6$ Step 2: $⅓ \times 12 = 4$ Step 3: $6 + 4 = 10$.

Hence $(½ \times 12) + (⅓ \times 12) = 10$.

ADDITIONAL SUGGESTIONS

Children find Picture Puzzles with Cuisenaire® Rods motivating and easy to use. Teased by the titles, they try to guess what the picture will be; sometimes it does not reveal itself until the last rod is placed. Once the rod picture is completed, any errors in answers become visually evident. This self-checking feature of the book, together with the use of the rod model to solve the problems in the first place, helps children strengthen basic computational skills. Coloring the rod pictures is optional; however, the coloring helps children know the rod lengths and colors, as well as strengthen fine motor development and spatial awareness.

Complete answers are given in the back of the book to show how each picture should look. The pages are reproducible, and children can choose them in whatever order fits their needs. The coverage is essentially the arithmetical topics of Grades K-4, yet the book is valuable for "drill and practice" and remediation work in the upper grades. Older students enjoy doing the problems in their heads if they can, using the rod model to solve only some of the problems. Because they need the rods to make the pictures, they are not embarrassed to have rods on their desks to do this review and remedial work.

Additional problems can be made for the same rod pictures. Duplicate the gridded half of the sheet, observe the answers in the answer key, and make up new problems for these same answers. Children enjoy making their own problem sets, perhaps learning as much from making problems with given answers as from finding answers to given problems. Five Picture Puzzle Masters (pages 46-50) have been left open-ended for teachers or students to have fun making the problems (pages 46-48) and the problems and pictures (pages 49-50). Please share your Picture Puzzles with us.

A Big _____

i		o		n	h		m			k			b	
												j		l
	c				d									
				g				a			e		f	

Rule: White = 1

ACROSS

a) $1 + 1 =$ _____
b) $2 + 1 =$ _____
c) $2 - 1 =$ _____
d) $3 - 2 =$ _____
e) $5 - 3 =$ _____
f) $1 + 2 =$ _____
g) $4 - 1 =$ _____
h) $0 + 2 =$ _____

DOWN

i) $3 + 2 =$ _____
j) $5 - 2 =$ _____
k) $4 + 1 =$ _____
l) $3 + 0 =$ _____
m) $1 + 4 =$ _____
n) $2 + 2 =$ _____
o) $5 + 0 =$ _____

ROD WORK AREA

The Open Road

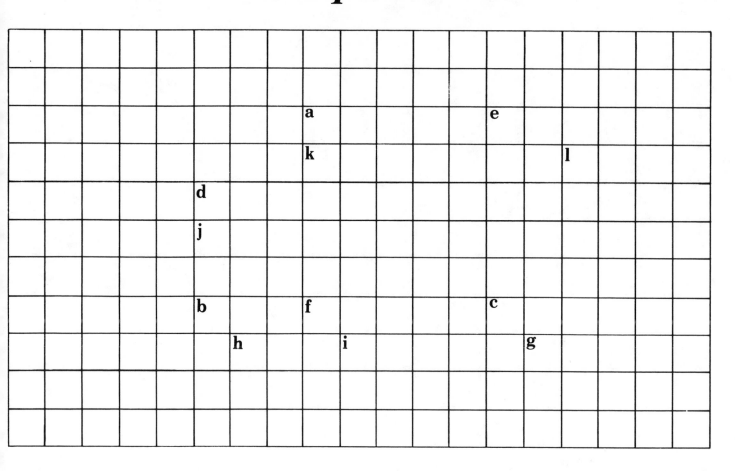

Rule: White = 1

ACROSS

a) 2 + 3 = _____
b) 3 + 0 = _____
c) 4 − 1 = _____
d) 5 − 2 = _____
e) 2 + 1 = _____
f) 1 + 4 = _____

DOWN

g) 5 − 4 = _____
h) 1 + 0 = _____
i) 3 − 2 = _____
j) 0 + 2 = _____
k) 5 − 1 = _____
l) 2 + 2 = _____

ROD WORK AREA

7

Picture Puzzles with Cuisenaire Rods
© 1979 Cuisenaire Company of America, Inc.

Mirror Magic

Rule: White = 1

ACROSS

a) 3 + 0 = ____
b) 2 + 2 = ____
c) 5 − 2 = ____
d) 1 + 3 = ____
e) 5 − 1 = ____
f) 2 + 1 = ____
g) 3 + 1 = ____
h) 4 − 1 = ____

DOWN

i) 2 + 3 = ____
j) 4 − 2 = ____
k) 0 + 5 = ____
l) 1 + 1 = ____
m) 1 + 4 = ____
n) 3 + 2 = ____
o) 5 − 3 = ____
p) 3 − 1 = ____

ROD WORK AREA

Discovering America

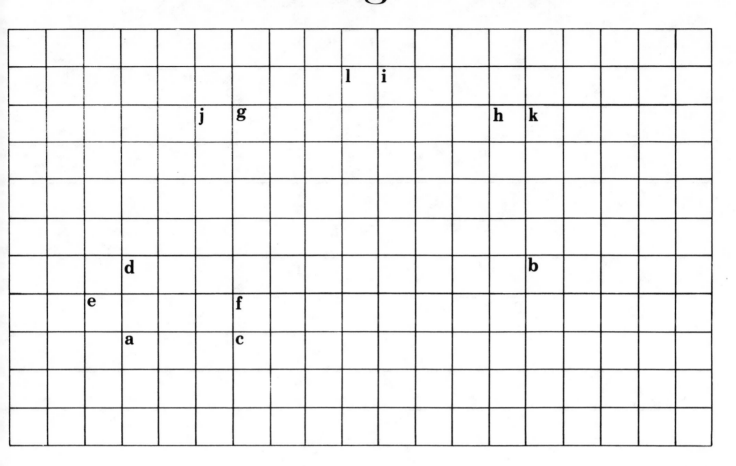

Rule: White = 1

ACROSS

a) $7 - 4 =$ _____
b) $8 - 6 =$ _____
c) $5 + 4 =$ _____
d) $9 - 5 =$ _____
e) $6 - 2 =$ _____
f) $6 + 3 =$ _____

DOWN

g) $6 - 5 =$ _____
h) $10 - 5 =$ _____
i) $9 - 8 =$ _____
j) $7 - 3 =$ _____
k) $7 - 6 =$ _____
l) $3 + 3 =$ _____

ROD WORK AREA

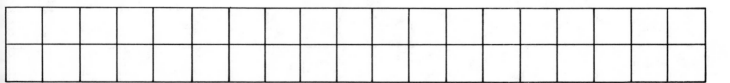

Picture Puzzles with Cuisenaire Rods
© 1979 Cuisenaire Company of America, Inc.

The Great _____

[Grid puzzle with labeled cells: k, b, i, c, f, d, a, m, j, h, l, g, e]

Rule: White = 1

ACROSS

a) $1 + 0 =$ ____

b) $6 + 4 =$ ____

c) $10 - 8 =$ ____

d) $6 - 1 =$ ____

e) $7 + 3 =$ ____

f) $9 - 7 =$ ____

g) $9 - 4 =$ ____

DOWN

h) $7 - 6 =$ ____

i) $5 + 5 =$ ____

j) $8 - 7 =$ ____

k) $2 + 8 =$ ____

l) $4 - 3 =$ ____

m) $10 - 9 =$ ____

ROD WORK AREA

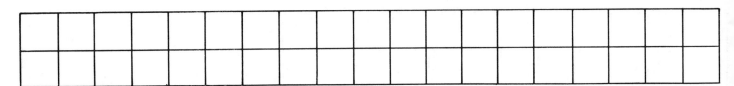

My Teacher

Rule: White = 1

ACROSS

a)
$$\begin{array}{r} 6 \\ -\ 4 \\ \hline \end{array}$$

b)
$$\begin{array}{r} 2 \\ +\ 1 \\ \hline \end{array}$$

c)
$$\begin{array}{r} 7 \\ -\ 5 \\ \hline \end{array}$$

d)
$$\begin{array}{r} 3 \\ +\ 0 \\ \hline \end{array}$$

e)
$$\begin{array}{r} 8 \\ -\ 6 \\ \hline \end{array}$$

f)
$$\begin{array}{r} 8 \\ -\ 5 \\ \hline \end{array}$$

g)
$$\begin{array}{r} 0 \\ +\ 2 \\ \hline \end{array}$$

DOWN

h)
$$\begin{array}{r} 4 \\ +\ 3 \\ \hline \end{array}$$

i)
$$\begin{array}{r} 10 \\ -\ 7 \\ \hline \end{array}$$

j)
$$\begin{array}{r} 4 \\ +\ 4 \\ \hline \end{array}$$

k)
$$\begin{array}{r} 4 \\ -\ 1 \\ \hline \end{array}$$

l)
$$\begin{array}{r} 2 \\ +\ 5 \\ \hline \end{array}$$

m)
$$\begin{array}{r} 6 \\ -\ 3 \\ \hline \end{array}$$

n)
$$\begin{array}{r} 3 \\ +\ 5 \\ \hline \end{array}$$

o)
$$\begin{array}{r} 10 \\ -\ 8 \\ \hline \end{array}$$

ROD WORK AREA

Picture Puzzles with Cuisenaire Rods
© 1979 Cuisenaire Company of America, Inc.

Pin-Wheel Pattern

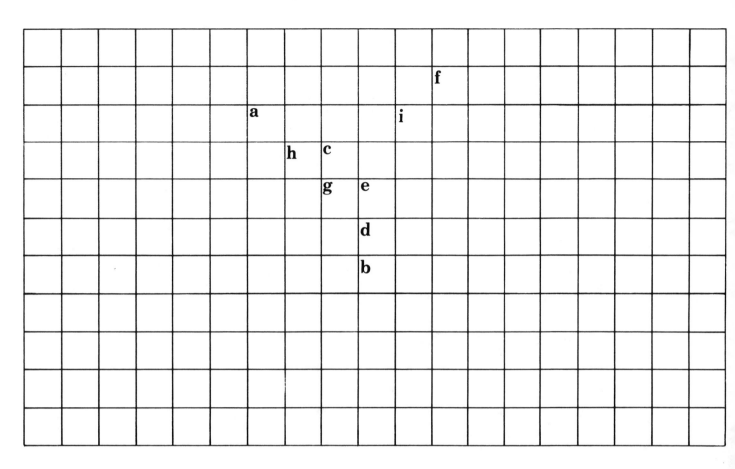

Rule: White = 1

ACROSS

a) _____ is one more than 3

b) _____ is one less than 5

c) _____ is one more than 1

d) _____ is one less than 3

DOWN

e) _____ is one less than 2

f) _____ is one more than 4

g) _____ is one less than 4

h) _____ is one less than 6

i) _____ is one more than 2

ROD WORK AREA

One To Eight

Rule: White = 1

ACROSS		DOWN	
a)	__ is between 5 and 7	i)	__ is between 8 and 6
b)	__ is between 4 and 2	j)	__ is between 1 and 3
c)	__ is between 7 and 9	k)	__ is between 3 and 5
d)	__ is between 4 and 6	l)	__ is between 2 and 0
e)	__ is between 0 and 2	m)	__ is between 9 and 7
f)	__ is between 5 and 3	n)	__ is between 7 and 5
g)	__ is between 3 and 1	o)	__ is between 2 and 4
h)	__ is between 6 and 8	p)	__ is between 6 and 4

ROD WORK AREA

13

I'm Going _____

Rule: White = 1

Which is Greater Than?

ACROSS		DOWN	
a) 2 or 3 _____		h) 8 or 6 _____	
b) 4 or 2 _____		i) 4 or 3 _____	
c) 4 or 5 _____		j) 9 or 8 _____	
d) 3 or 1 _____		k) 7 or 8 _____	
e) 0 or 2 _____		l) 7 or 9 _____	
f) 3 or 4 _____			
g) 3 or 0 _____			

ROD WORK AREA

Desert Friend

Rule: White = 1

Which is Less?

ACROSS

a) 3 or 6 _____

b) 2 or 4 _____

c) 6 or 5 _____

d) 10 or 8 _____

e) 1 or 2 _____

f) 5 or 3 _____

g) 7 or 8 _____

DOWN

h) 4 or 6 _____

i) 3 or 2 _____

j) 4 or 5 _____

k) 2 or 1 _____

l) 1 or 3 _____

m) 1 or 2 _____

ROD WORK AREA

Picture Puzzles with Cuisenaire Rods

One Step At A Time

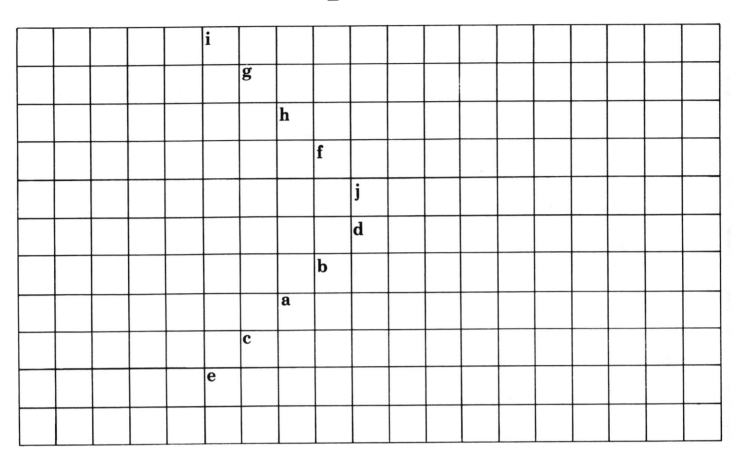

Rule: White = 1

ACROSS

a) 1, 2, 3, 4, 5, __, 7, 8
b) 1, 2, 3, __, 5, 6
c) 10, 9, __, 7, 6
d) 5, 4, 3, __, 1
e) 2, 4, 6, 8, ___, 12, 14

DOWN

f) 6, 5, 4, __, 2, 1
g) 1, 3, 5, __, 9, 11
h) __, 10, 15, 20, 25
i) 3, 6, __, 12, 15
j) 7, 5, 3, __

ROD WORK AREA

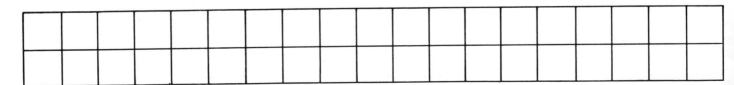

Are You My Mother?

Rule: White = 1
Making "Orange Plus" Trains

ACROSS

a) 13, 14, 15, 1__, 17, 18
b) 15, 14, 1__, 12, 11
c) 10, 1__, 14, 16
d) 10, 1__, 12, 13
e) 18, 16, 14, 1__, 10
f) 19, 18, 17, 1__, 15, 14

DOWN

g) 18, 20, 2__, 24, 26
h) 25, 24, 2__, 22, 21
i) 10, 1__, 14, 16
j) 10, 15, 20, 2__, 30
k) 21, 20, 1__, 18, 17
l) 23, 2__, 19, 17

ROD WORK AREA

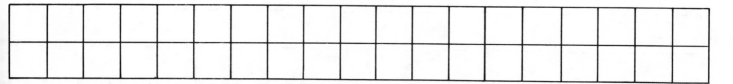

Picture Puzzles with Cuisenaire Rods
© 1979 Cuisenaire Company of America, Inc.

Under Water

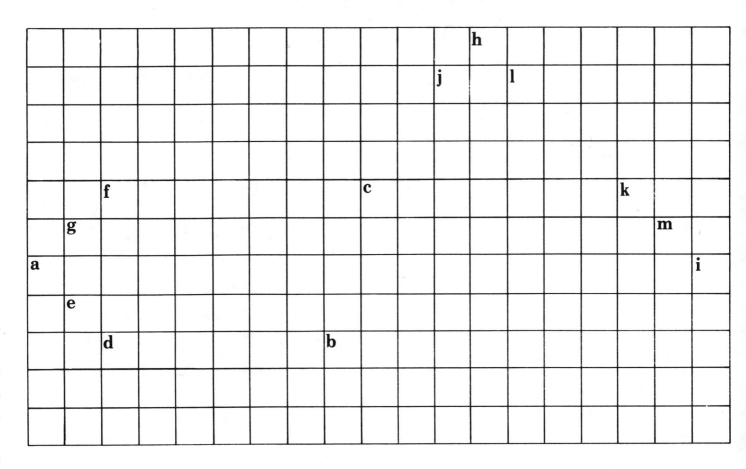

Rule: White = 1

ACROSS

a) 24 = 2 tens + _____ ones
b) 18 = 1 ten + _____ ones
c) 37 = 3 tens + _____ ones
d) 16 = 1 ten + _____ ones
e) 33 = 3 tens + _____ ones
f) 17 = 1 ten + _____ ones
g) 23 = 2 tens + _____ ones

DOWN

h) 12 = 1 ten + _____ ones
i) 21 = 2 tens + _____ ones
j) 13 = 1 ten + _____ one
k) 25 = 2 tens + _____ ones
l) 43 = 4 tens + _____ ones
m)   3 = 0 tens + _____ ones

ROD WORK AREA

tens	ones

Circus Fun

ACROSS

a) 2 tens + 5 ones = 2____
b) 1 ten + 7 ones = 1____
c) 2 tens + 1 one = 2____
d) 2 tens + 3 ones = 2____
e) 1 ten + 9 ones = 1____
f) 1 ten + 3 ones = 1____

DOWN

g) 3 tens + 2 ones = 3____
h) 2 tens + 4 ones = 2____
i) 1 ten + 2 ones = 1____
j) 1 ten + 4 ones = 1____
k) 3 tens + 1 one = 3____
l) 2 tens + 2 ones = 2____

ROD WORK AREA

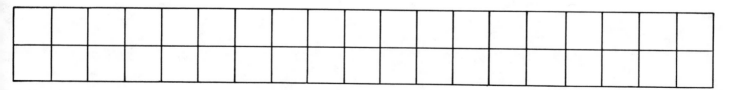

Picture Puzzles with Cuisenaire Rods
© 1979 Cuisenaire Company of America, Inc.

Season's Greetings

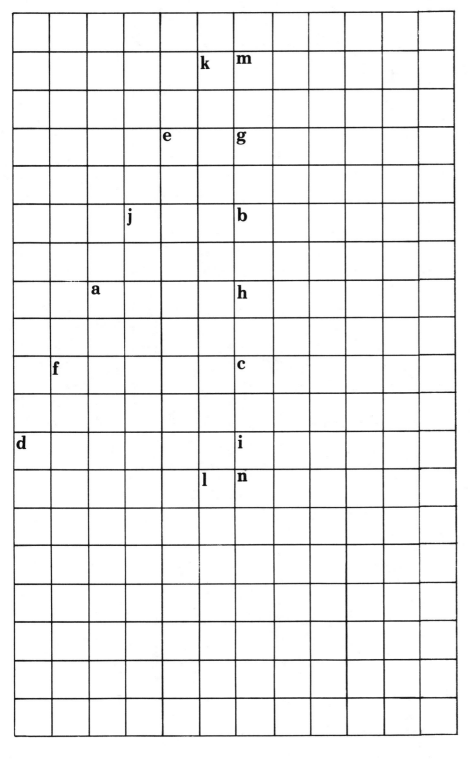

Rule: White = 1

ACROSS

a) 2 + ___ = 6
b) 7 + ___ = 10
c) 3 + ___ = 8
d) 1 + ___ = 7
e) 0 + ___ = 2
f) 5 + ___ = 10
g) 7 + ___ = 9
h) 4 + ___ = 8
i) 3 + ___ = 9
j) 5 + ___ = 8

DOWN

k) 9 + ___ = 10
l) 6 + ___ = 8
m) 7 + ___ = 8
n) 4 + ___ = 6

ROD WORK AREA

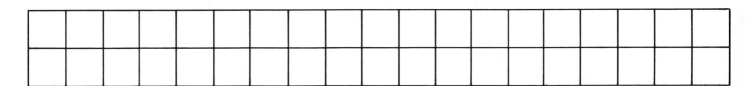

To The Rescue

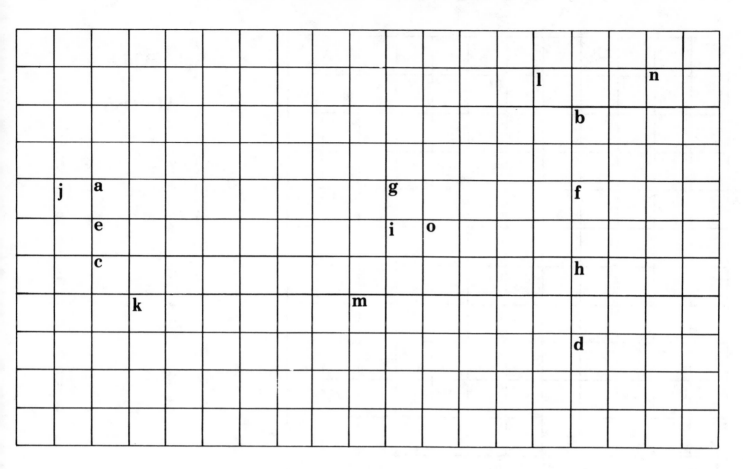

Rule: White = 1

ACROSS

a) ____ + 2 = 10
b) ____ + 4 = 6
c) ____ + 1 = 9
d) ____ + 6 = 8
e) ____ + 0 = 8
f) ____ + 8 = 10
g) ____ + 5 = 6
h) ____ + 2 = 4

DOWN

i) ____ + 5 = 7
j) ____ + 6 = 9
k) ____ + 7 = 8
l) ____ + 1 = 10
m) ____ + 9 = 10
n) ____ + 0 = 9
o) ____ + 3 = 5

ROD WORK AREA

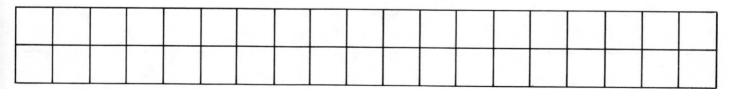

Desert Flower

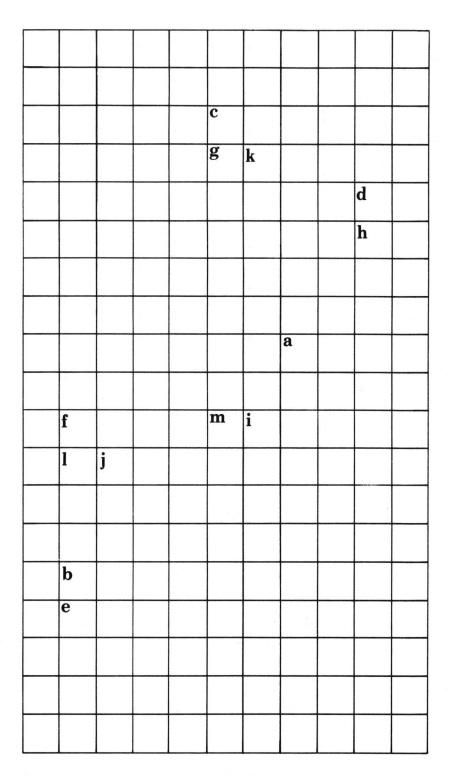

Rule: White = 1

ACROSS

a) $4 + \underline{\quad} = 7$

b) $\underline{\quad} + 2 = 6$

c) $3 + \underline{\quad} = 5$

d) $0 + \underline{\quad} = 1$

e) $\underline{\quad} + 6 = 10$

f) $\underline{\quad} + 8 = 10$

DOWN

g) $2 + \underline{\quad} = 9$

h) $\underline{\quad} + 2 = 5$

i) $\underline{\quad} + 1 = 10$

j) $5 + \underline{\quad} = 8$

k) $1 + \underline{\quad} = 8$

l) $3 + \underline{\quad} = 6$

m) $\underline{\quad} + 0 = 9$

ROD WORK AREA

Sssss

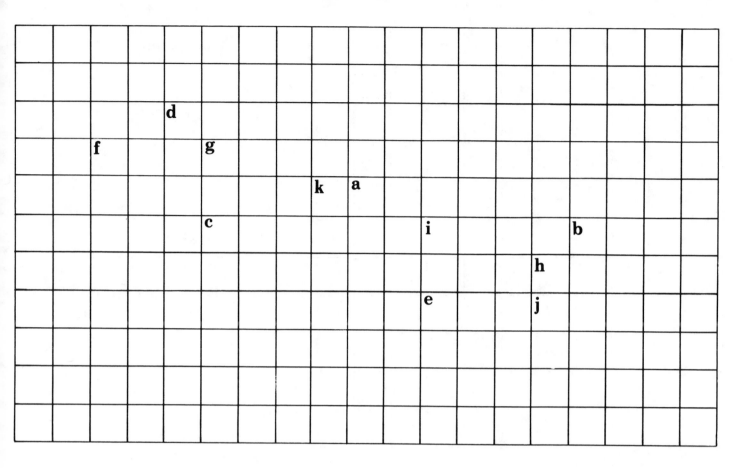

Rule: White = 1

ACROSS

a) 8 − ____ = 5
b) 6 − ____ = 5
c) 4 − ____ = 1
d) 4 − ____ = 3
e) 9 − ____ = 6
f) 7 − ____ = 4

DOWN

g) 9 − ____ = 7
h) 7 − ____ = 6
i) 6 − ____ = 4
j) 9 − ____ = 8
k) 10 − ____ = 8

ROD WORK AREA

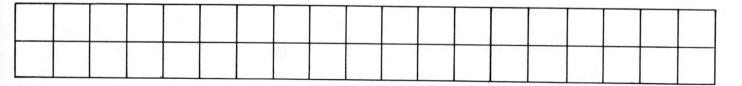

More Mirror Magic

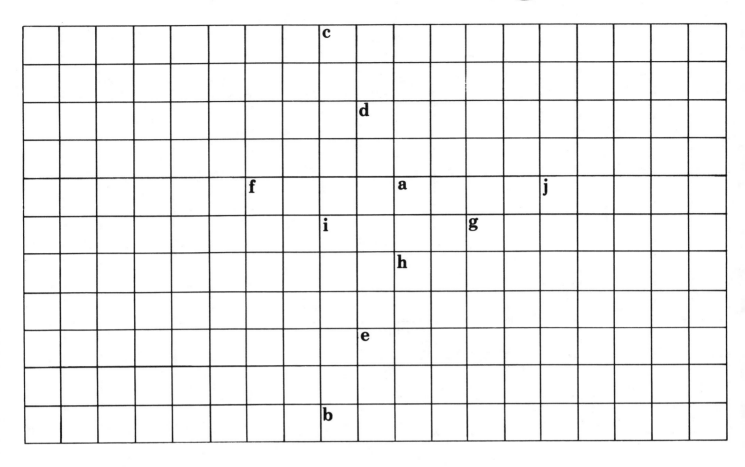

Rule: White = 1

ACROSS

a) 10 − 9 = ____
b) 9 − ____ = 4
c) 4 + ____ = 9
d) 7 + ____ = 10
e) 7 − 4 = ____

DOWN

f) 6 + ____ = 10
g) 6 − ____ = 4
h) 8 − 7 = ____
i) 10 − 8 = ____
j) 5 + ____ = 9

ROD WORK AREA

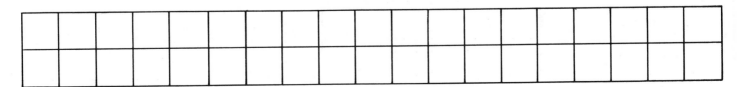

Mr. Robot

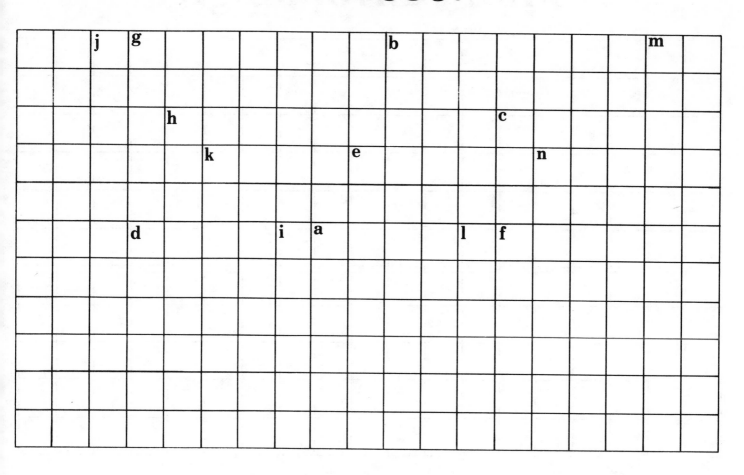

Rule: White = 1

ACROSS

a) $2 + 3 + \underline{\hphantom{00}} = 9$

b) $10 - 3 = \underline{\hphantom{00}}$

c) $\underline{\hphantom{00}} + 4 + 1 = 8$

d) $2 + \underline{\hphantom{00}} + 4 = 10$

e) $\underline{\hphantom{00}} = 7 - 5$

f) $10 - 6 = \underline{\hphantom{00}}$

g) $\underline{\hphantom{00}} = 2 + 1 + 4$

h) $3 + \underline{\hphantom{00}} + 3 = 9$

DOWN

i) $5 = 10 - \underline{\hphantom{00}}$

j) $\underline{\hphantom{00}} = 1 + 2 + 3$

k) $4 + \underline{\hphantom{00}} + 5 = 10$

l) $9 - \underline{\hphantom{00}} = 4$

m) $4 = 10 - \underline{\hphantom{00}}$

n) $2 + \underline{\hphantom{00}} + 1 + 3 = 7$

ROD WORK AREA

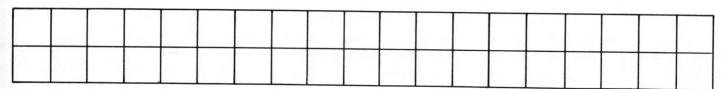

25

Riding In Style

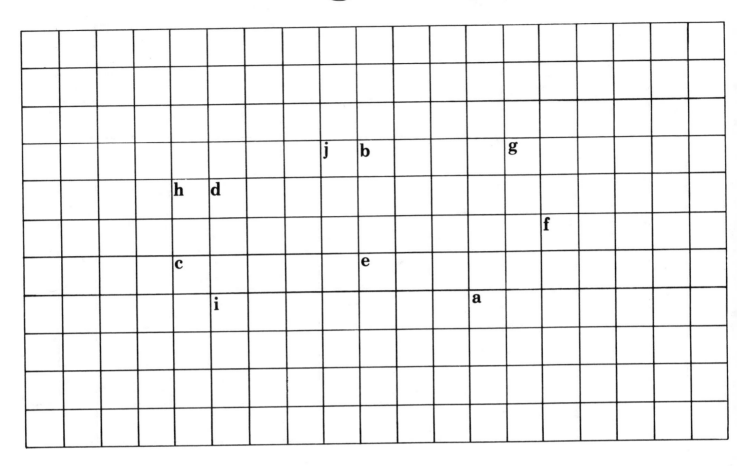

Rule: White = 1
Making "Orange Plus" Trains

ACROSS	DOWN
a) $7 + 4 = 10 + \underline{\quad}$	**f)** $9 + 3 = 10 + \underline{\quad}$
b) $8 + 6 = 10 + \underline{\quad}$	**g)** $5 + 8 = 10 + \underline{\quad}$
c) $8 + 7 = 10 + \underline{\quad}$	**h)** $7 + 5 = 10 + \underline{\quad}$
d) $6 + 7 = 10 + \underline{\quad}$	**i)** $9 + 2 = 10 + \underline{\quad}$
e) $9 + 6 = 10 + \underline{\quad}$	**j)** $6 + 6 = 10 + \underline{\quad}$

ROD WORK AREA

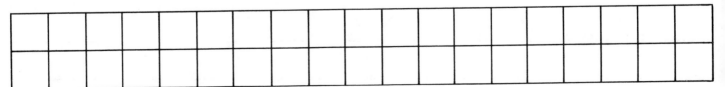

What's For Dinner?

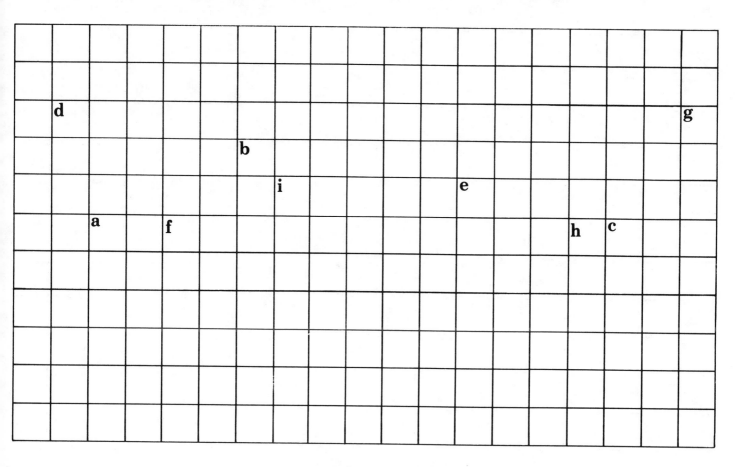

Rule: White = 1
Making "Orange Plus" Trains

ACROSS

a) $5 + 6 + 1 = 10 +$ ____
b) $4 + 5 + 9 = 10 +$ ____
c) $3 + 5 + 4 = 10 +$ ____

DOWN

d) $7 + 4 + 5 = 10 +$ ____
e) $2 + 9 + 3 = 10 +$ ____
f) $3 + 6 + 4 = 10 +$ ____
g) $8 + 2 + 6 = 10 +$ ____
h) $4 + 7 + 2 = 10 +$ ____
i) $1 + 9 + 4 = 10 +$ ____

ROD WORK AREA

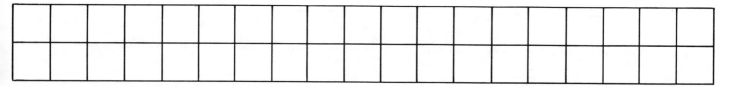

Upside Down

Rule: White = 1

ACROSS

a)
$$\begin{array}{r} 2\,1 \\ +1\,1 \\ \hline 3\,\underline{} \end{array}$$

b)
$$\begin{array}{r} 1\,1 \\ +1\,3 \\ \hline 2\,\underline{} \end{array}$$

c)
$$\begin{array}{r} 1\,3 \\ +1\,1 \\ \hline 2\,\underline{} \end{array}$$

d)
$$\begin{array}{r} 1\,1 \\ +2\,1 \\ \hline 3\,\underline{} \end{array}$$

e)
$$\begin{array}{r} 3\,3 \\ +1\,1 \\ \hline 4\,\underline{} \end{array}$$

f)
$$\begin{array}{r} 1\,1 \\ +3\,3 \\ \hline 4\,\underline{} \end{array}$$

g)
$$\begin{array}{r} 2\,1 \\ +3\,1 \\ \hline 5\,\underline{} \end{array}$$

h)
$$\begin{array}{r} 3\,1 \\ +2\,1 \\ \hline 5\,\underline{} \end{array}$$

DOWN

i)
$$\begin{array}{r} 2\,3 \\ +1\,4 \\ \hline 3\,\underline{} \end{array}$$

j)
$$\begin{array}{r} 4\,2 \\ +3\,1 \\ \hline 7\,\underline{} \end{array}$$

k)
$$\begin{array}{r} 3\,1 \\ +4\,2 \\ \hline 7\,\underline{} \end{array}$$

l)
$$\begin{array}{r} 1\,4 \\ +2\,3 \\ \hline 3\,\underline{} \end{array}$$

m)
$$\begin{array}{r} 5\,0 \\ +2\,3 \\ \hline 7\,\underline{} \end{array}$$

n)
$$\begin{array}{r} 2\,3 \\ +5\,0 \\ \hline 7\,\underline{} \end{array}$$

ROD WORK AREA

tens	ones

Gee Whiz

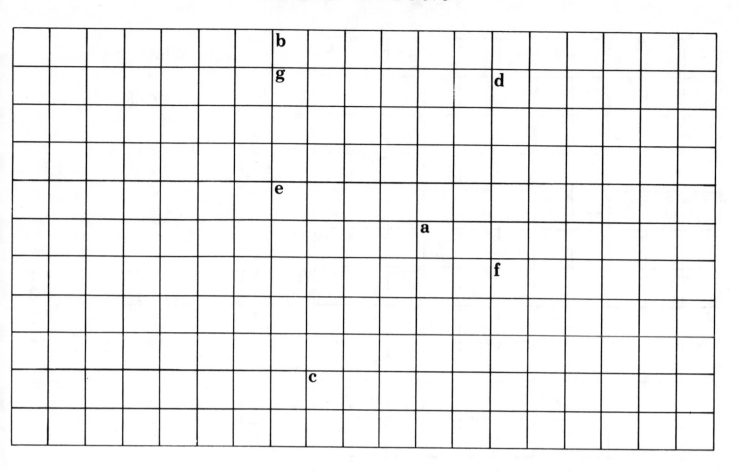

Rule: White = 1

ACROSS

a) $\begin{array}{r} 27 \\ +16 \\ \hline 4\_\_ \end{array}$
b) $\begin{array}{r} 19 \\ +18 \\ \hline 3\_\_ \end{array}$

c) $\begin{array}{r} 16 \\ +29 \\ \hline 4\_\_ \end{array}$

DOWN

d) $\begin{array}{r} 35 \\ +17 \\ \hline 5\_\_ \end{array}$
e) $\begin{array}{r} 18 \\ +18 \\ \hline 3\_\_ \end{array}$
f) $\begin{array}{r} 27 \\ +17 \\ \hline 4\_\_ \end{array}$

g) $\begin{array}{r} 18 \\ +25 \\ \hline 4\_\_ \end{array}$

ROD WORK AREA

tens	ones

29

Stretching-Up

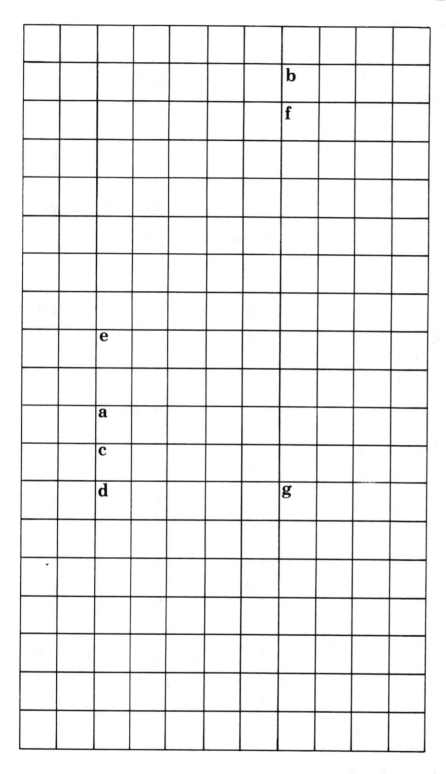

Rule: White = 1

ACROSS

a) $12 - 7 = $ ____
b) $13 - 9 = $ ____
c) $14 - 9 = $ ____

DOWN

d) $14 - 8 = $ ____
e) $11 - 9 = $ ____
f) $20 - 10 = $ ____
g) $13 - 7 = $ ____

ROD WORK AREA

Goofy

ACROSS

a) $\begin{array}{r} 26 \\ -14 \\ \hline 1\_\_ \end{array}$ b) $\begin{array}{r} 39 \\ -14 \\ \hline 2\_\_ \end{array}$ c) $\begin{array}{r} 48 \\ -15 \\ \hline 3\_\_ \end{array}$

d) $\begin{array}{r} 37 \\ -25 \\ \hline 1\_\_ \end{array}$ e) $\begin{array}{r} 28 \\ -14 \\ \hline 1\_\_ \end{array}$ f) $\begin{array}{r} 49 \\ -17 \\ \hline 3\_\_ \end{array}$

g) $\begin{array}{r} 58 \\ -13 \\ \hline 4\_\_ \end{array}$ h) $\begin{array}{r} 27 \\ -16 \\ \hline 1\_\_ \end{array}$ i) $\begin{array}{r} 49 \\ -26 \\ \hline 2\_\_ \end{array}$

DOWN

j) $\begin{array}{r} 48 \\ -20 \\ \hline 2\_\_ \end{array}$ k) $\begin{array}{r} 29 \\ -18 \\ \hline 1\_\_ \end{array}$ l) $\begin{array}{r} 57 \\ -24 \\ \hline 3\_\_ \end{array}$

m) $\begin{array}{r} 76 \\ -14 \\ \hline 6\_\_ \end{array}$ n) $\begin{array}{r} 59 \\ -23 \\ \hline 3\_\_ \end{array}$ o) $\begin{array}{r} 79 \\ -21 \\ \hline 5\_\_ \end{array}$

p) $\begin{array}{r} 98 \\ -22 \\ \hline 7\_\_ \end{array}$ q) $\begin{array}{r} 85 \\ -32 \\ \hline 5\_\_ \end{array}$ r) $\begin{array}{r} 99 \\ -67 \\ \hline 3\_\_ \end{array}$

ROD WORK AREA

tens	ones

Go Fly A _____

Rule: White = 1

ACROSS

a)   21
    −19
    ——

b)   23
    −17
    ——

c)   30
    −29
    ——

d)   34
    −28
    ——

e)   30
    −28
    ——

DOWN

f)   35
    −19
    1_

g)   40
    −28
    1_

h)   50
    −24
    2_

i)   41
    −19
    2_

j)   60
    −48
    1_

ROD WORK AREA

tens	ones

The Big Push

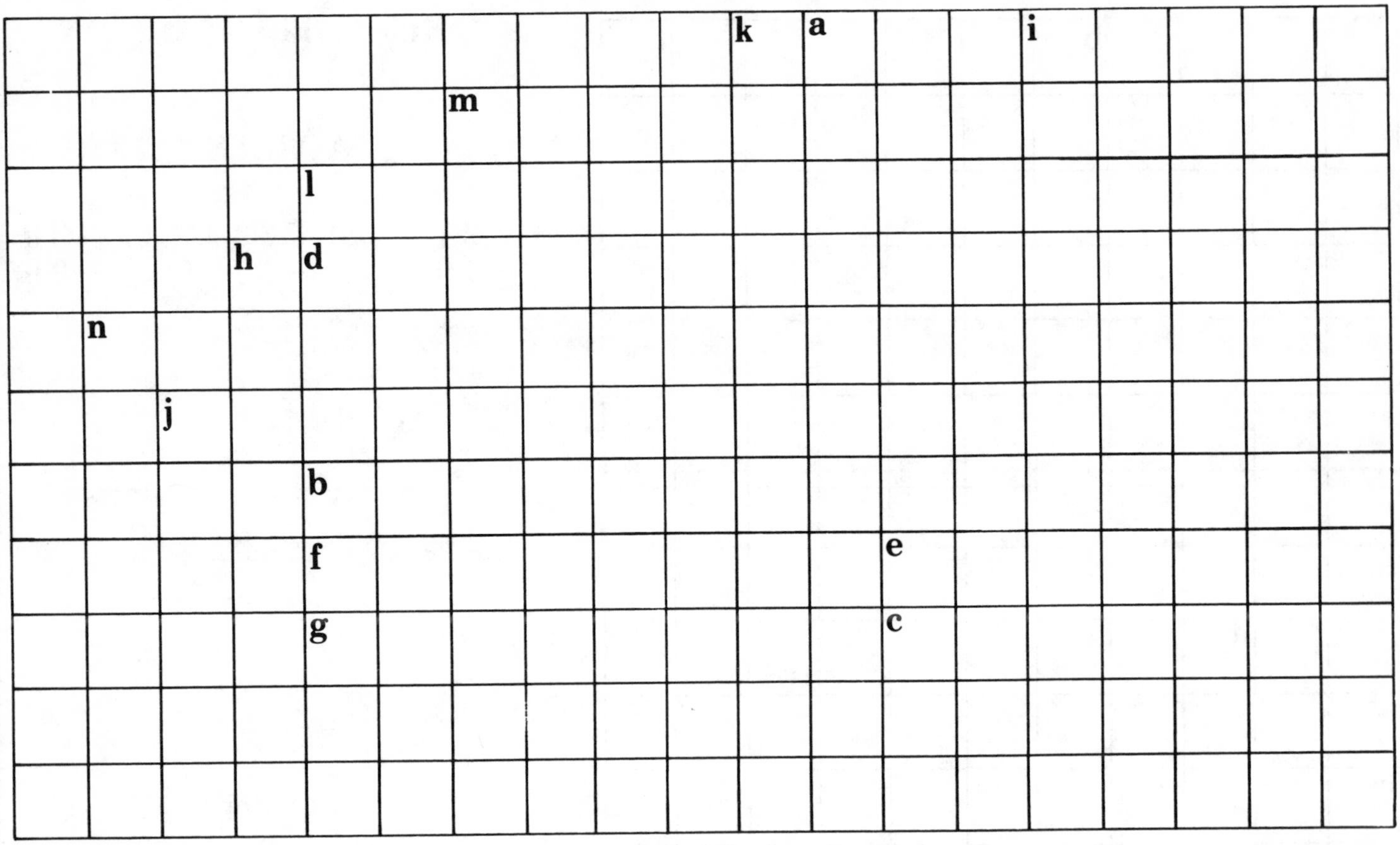

(grid puzzle with labeled cells: k, a, i across the top; m, l, h, d, n, j, b, f, e, g, c)

Rule: White = 1

ACROSS

a)
```
  21
+32
 5_
```

b)
```
  50
-40
___
```

c)
```
  29
-17
 1_
```

d)
```
  25
-15
___
```

e)
```
  29
+13
 4_
```

f)
```
  76
-34
 4_
```

g)
```
  17
+25
 4_
```

DOWN

h)
```
  29
+25
 5_
```

i)
```
  49
-12
 3_
```

j)
```
  50
+32
 8_
```

k)
```
  17
+26
 4_
```

l)
```
  49
-28
 2_
```

m)
```
  38
+24
 6_
```

n)
```
  54
-39
 1_
```

ROD WORK AREA

tens	ones

Picture Puzzles with Cuisenaire Rods
© 1979 Cuisenaire Company of America, Inc.

Square Deal

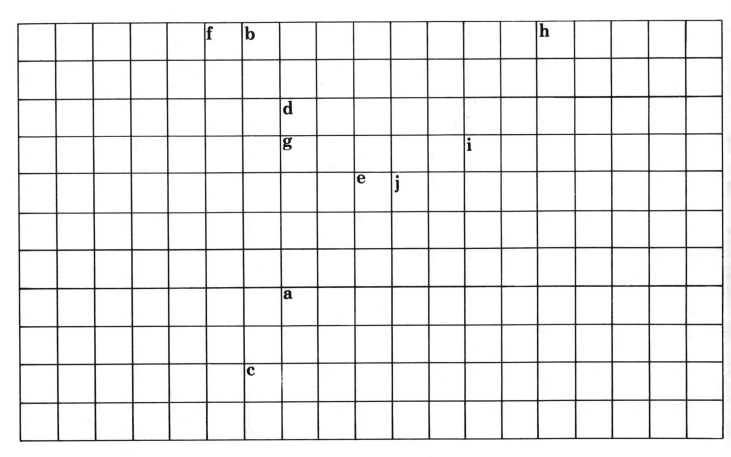

Rule: White = 1

ACROSS

a) $2 \times 3 =$ _____
b) $4 \times 2 =$ _____
c) $1 \times 8 =$ _____
d) $3 \times 2 =$ _____

DOWN

e) $1 \times 2 =$ _____
f) $2 \times 5 =$ _____
g) $2 \times 2 =$ _____
h) $10 \times 1 =$ _____
i) $1 \times 4 =$ _____
j) $2 \times 1 =$ _____

ROD WORK AREA

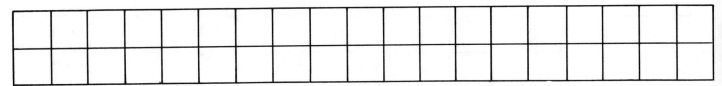

_____, *Rhymes With Noodle*

			c												
		j	k			n	m								
			g		d										
			p										a		
										l	f				
			b												
			h												
			e												
			i						o						

Rule: White = 1

ACROSS

a) $7 \times 3 = 20 +$ _____
b) $7 \times 7 = 40 +$ _____
c) $9 \times 5 = 40 +$ _____
d) $9 \times 9 = 80 +$ _____
e) $3 \times 13 = 30 +$ _____
f) $3 \times 7 = 20 +$ _____
g) $1 \times 21 = 20 +$ _____
h) $1 \times 11 = 10 +$ _____

DOWN

i) $4 \times 8 = 30 +$ _____
j) $6 \times 7 = 40 +$ _____
k) $8 \times 8 = 60 +$ _____
l) $6 \times 4 = 20 +$ _____
m) $11 \times 2 = 20 +$ _____
n) $9 \times 6 = 50 +$ _____
o) $4 \times 3 = 10 +$ _____
p) $4 \times 8 = 30 +$ _____

ROD WORK AREA

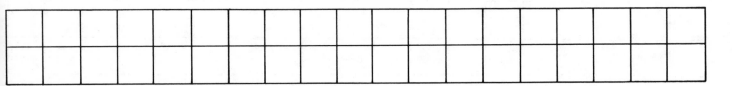

_____-Time

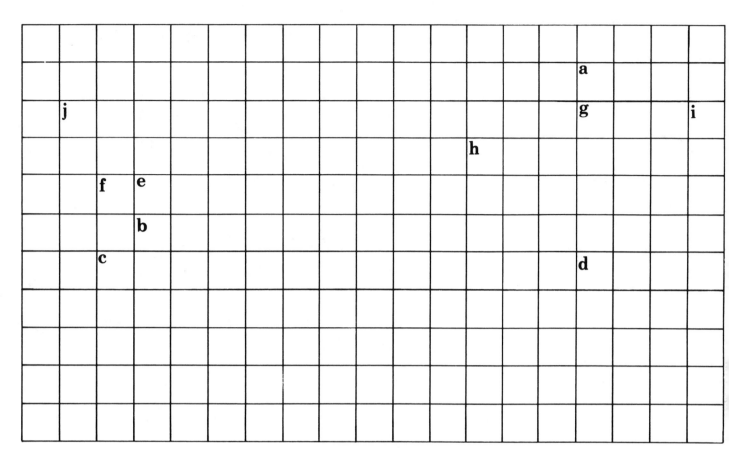

Rule: White = 1

ACROSS

a) $(2 \times 8) - (3 \times 4) = $ ____
b) $(5 \times 3) - (3 \times 2) = $ ____
c) $(6 \times 3) - (2 \times 4) = $ ____
d) $(4 \times 4) - (6 \times 2) = $ ____
e) $(8 \times 2) - (3 \times 5) = $ ____

DOWN

f) $(4 \times 3) - (5 \times 2) = $ ____
g) $(5 \times 4) - (4 \times 4) = $ ____
h) $(3 \times 8) - (6 \times 3) = $ ____
i) $(6 \times 4) - (4 \times 5) = $ ____
j) $(7 \times 3) - (2 \times 7) = $ ____

ROD WORK AREA

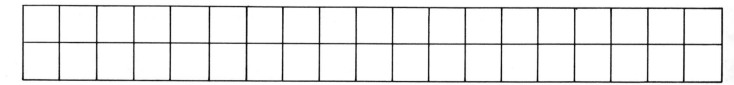

You Name It

Rule: White = 1

ACROSS

a)   1 3
  ×   3
  3 __

b)   2 1
  ×   3
  6 __

c)   1 1
  ×   7
  7 __

d)   1 7
  ×   3
  5 __

e)   1 1
  ×   5
  5 __

f)   1 7
  ×   1
  1 __

DOWN

g)   1 4
  ×   3
  4 __

h)   1 8
  ×   4
  7 __

i)   1 2
  ×   3
  3 __

j)   2 1
  ×   2
  4 __

k)   2 2
  ×   2
  4 __

l)   1 3
  ×   4
  5 __

ROD WORK AREA

tens	ones

Animal World

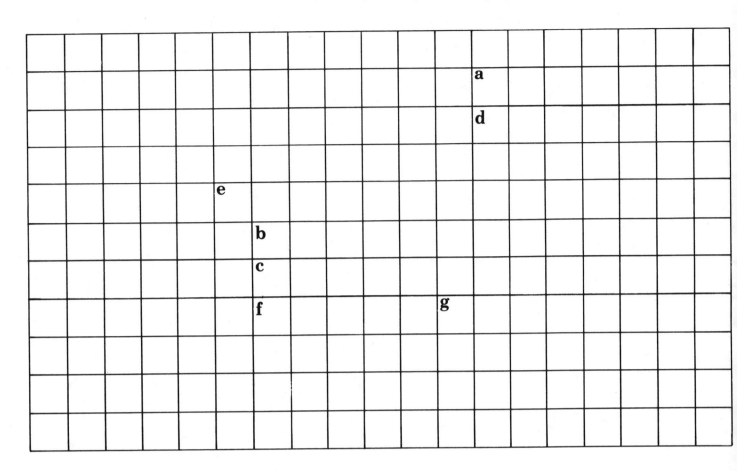

Rule: White = 1

ACROSS

a) ___ × 5 = 15
b) ___ × 3 = 18
c) ___ × 8 = 48

DOWN

d) ___ × 9 = 36
e) ___ × 13 = 13
f) ___ × 12 = 24
g) ___ × 9 = 18

ROD WORK AREA

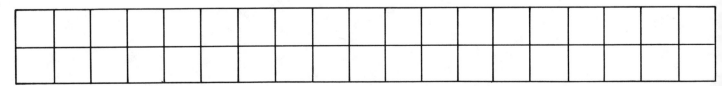

Teacher's Pet

		c						g								
			j	i		k				l	h		m			
			d		e					b						
			a						f							

Rule: White = 1

ACROSS

a) $14 \div 7 =$ _____
b) $16 \div 8 =$ _____
c) $21 \div 3 =$ _____
d) $8 \div 4 =$ _____
e) $16 \div 4 =$ _____
f) $12 \div 6 =$ _____
g) $28 \div 4 =$ _____
h) $10 \div 5 =$ _____
i) $18 \div 9 =$ _____

DOWN

j) $12 \div 2 =$ _____
k) $30 \div 5 =$ _____
l) $24 \div 4 =$ _____
m) $18 \div 3 =$ _____

ROD WORK AREA

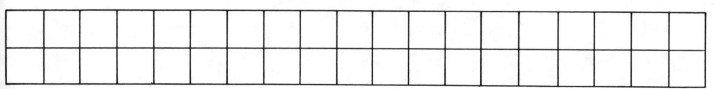

Turning Corners

Rule: White = 1

ACROSS

a) ½ × 4 = ____
b) ½ × 20 = ____
c) ½ × 12 = ____
d) ½ × 16 = ____
e) ½ × 8 = ____

DOWN

f) ½ × 18 = ____
g) ½ × 2 = ____
h) ½ × 10 = ____
i) ½ × 6 = ____
j) ½ × 14 = ____

ROD WORK AREA

Picture Puzzles with Cuisenaire Rods
© 1979 Cuisenaire Company of America, Inc.

A-Maz-ing

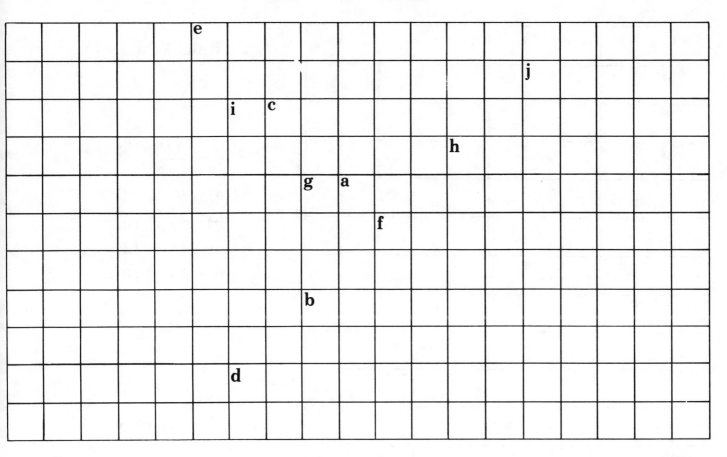

Rule: White = 1

ACROSS

a) $\frac{1}{3} \times 6 =$ _____
b) $\frac{1}{3} \times 12 =$ _____
c) $\frac{1}{3} \times 18 =$ _____
d) $\frac{1}{3} \times 24 =$ _____
e) $\frac{1}{3} \times 30 =$ _____

DOWN

f) $\frac{1}{3} \times 3 =$ _____
g) $\frac{1}{3} \times 9 =$ _____
h) $\frac{1}{3} \times 15 =$ _____
i) $\frac{1}{3} \times 21 =$ _____
j) $\frac{1}{3} \times 27 =$ _____

ROD WORK AREA

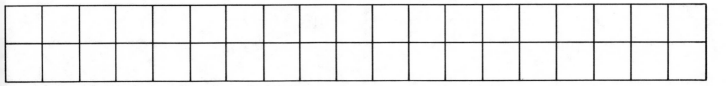

April Fools (Only If You Do Page 41)

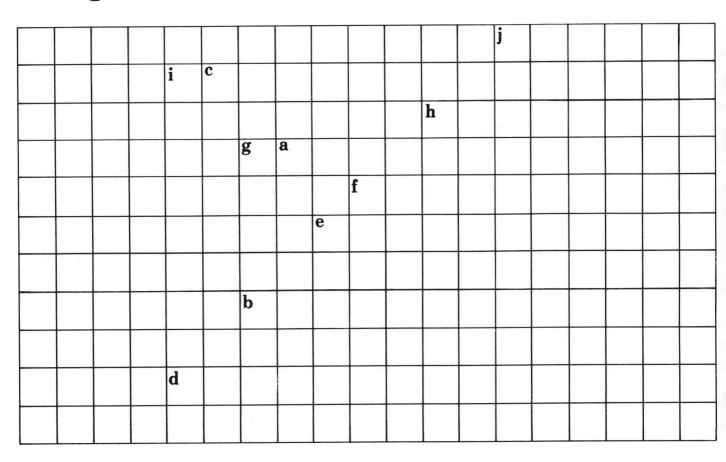

Rule: White = 1

ACROSS

a) ¼ × 12 = _____
b) ¼ × 20 = _____
c) ¼ × 28 = _____
d) ¼ × 36 = _____
e) ¼ × 4 = _____

DOWN

f) ¼ × 8 = _____
g) ¼ × 16 = _____
h) ¼ × 24 = _____
i) ¼ × 32 = _____
j) ¼ × 40 = _____

ROD WORK AREA

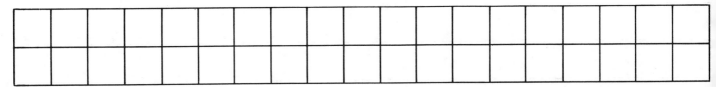

Puff, The Magic _____

Grid with labels: l, b, h, m, i, j, d, e, f, g, k, c, a

Rule: White = 1

ACROSS

a) $\frac{1}{2} \times 6 =$ ____
b) $\frac{1}{4} \times 8 =$ ____
c) $\frac{1}{2} \times 4 =$ ____
d) $\frac{1}{2} \times 2 =$ ____
e) $\frac{1}{3} \times 3 =$ ____
f) $\frac{1}{4} \times 4 =$ ____

DOWN

g) $\frac{1}{3} \times 6 =$ ____
h) $\frac{1}{3} \times 27 =$ ____
i) $\frac{1}{4} \times 32 =$ ____
j) $\frac{1}{4} \times 28 =$ ____
k) $6 \times \frac{1}{3} =$ ____
l) $30 \times \frac{1}{3} =$ ____
m) $8 \times \frac{1}{4} =$ ____

ROD WORK AREA

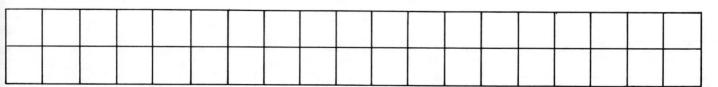

43

To And Fro

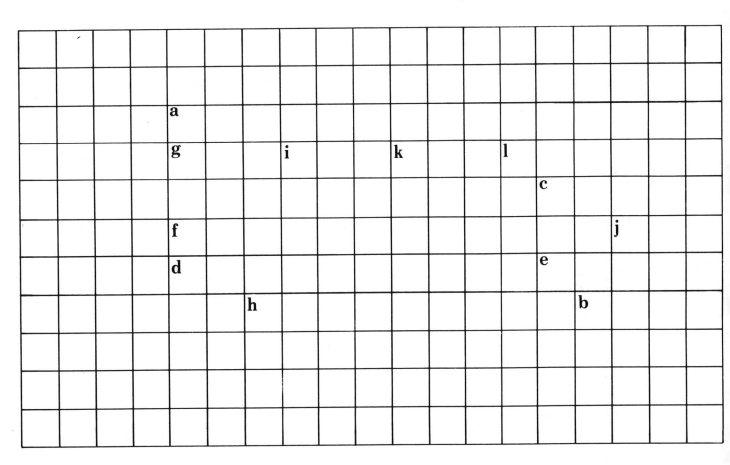

Rule: White = 1

ACROSS

a) $(\frac{1}{2} \times 12) + (\frac{1}{3} \times 12) =$ ____

b) $(\frac{1}{2} \times 1) + (\frac{1}{2} \times 1) =$ ____

c) $(\frac{1}{3} \times 6) + (\frac{1}{4} \times 4) =$ ____

d) $(\frac{1}{4} \times 20) + (\frac{1}{3} \times 15) =$ ____

e) $(\frac{1}{2} \times 2) + (\frac{1}{4} \times 8) =$ ____

f) $(\frac{1}{2} \times 14) + (\frac{1}{3} \times 9) =$ ____

DOWN

g) $(\frac{1}{4} \times 12) - (\frac{1}{3} \times 3) =$ ____

h) $(\frac{1}{4} \times 8) - (\frac{1}{3} \times 3) =$ ____

i) $(\frac{1}{3} \times 12) - (\frac{1}{3} \times 6) =$ ____

j) $(\frac{1}{2} \times 8) - (\frac{1}{4} \times 12) =$ ____

k) $(\frac{1}{4} \times 20) - (\frac{1}{2} \times 6) =$ ____

l) $(\frac{1}{3} \times 18) - (\frac{1}{4} \times 16) =$ ____

ROD WORK AREA

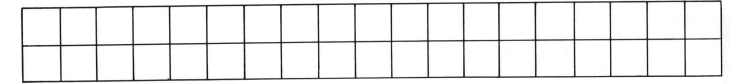

Be My Valentine

Rule: White = 1

ACROSS

a) $(\frac{1}{2} \times 6) - (\frac{1}{2} \times 2) =$ _____

b) $(\frac{1}{2} \times 2) + (\frac{1}{3} \times 3) =$ _____

c) $(\frac{1}{4} \times 12) - (\frac{1}{4} \times 4) =$ _____

d) $(\frac{1}{4} \times 4) + (\frac{1}{3} \times 3) =$ _____

e) $(\frac{1}{2} \times 12) - (\frac{1}{4} \times 16) =$ _____

f) $(\frac{1}{3} \times 12) - (\frac{1}{4} \times 8) =$ _____

g) $(\frac{1}{2} \times 10) - (\frac{1}{2} \times 6) =$ _____

h) $(\frac{1}{3} \times 12) - (\frac{1}{4} \times 12) =$ _____

DOWN

i) $(\frac{1}{3} \times 3) + (\frac{1}{4} \times 4) =$ _____

j) $(\frac{1}{4} \times 8) - (\frac{1}{3} \times 3) =$ _____

k) $(\frac{1}{3} \times 18) - (\frac{1}{3} \times 12) =$ _____

l) $(\frac{1}{4} \times 16) - (\frac{1}{4} \times 8) =$ _____

m) $(\frac{1}{3} \times 6) - (\frac{1}{4} \times 4) =$ _____

n) $(\frac{1}{2} \times 1) + (\frac{1}{2} \times 1) =$ _____

o) $(\frac{1}{3} \times 18) - (\frac{1}{4} \times 20) =$ _____

ROD WORK AREA

Picture Puzzles with Cuisenaire Rods
© 1979 Cuisenaire Company of America, Inc.

Once A Year

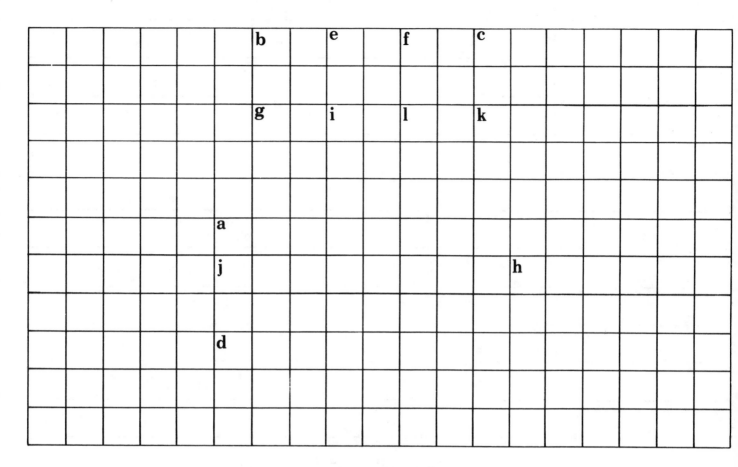

Rule: White = 1
Make Problems to Match the Answers

ACROSS		DOWN	
a)	= 9	g)	= 3
b)	= 1	h)	= 2
c)	= 1	i)	= 3
d)	= 9	j)	= 2
e)	= 1	k)	= 3
f)	= 1	l)	= 3

ROD WORK AREA

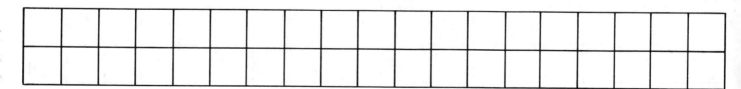

Quilt Pattern

Rule: White = 1
Make Problems to Match the Answers

ACROSS		DOWN	
a)	= 8	j)	= 8
b)	= 1	k)	= 4
c)	= 6	l)	= 2
d)	= 4	m)	= 6
e)	= 2	n)	= 6
f)	= 8	o)	= 4
g)	= 2	p)	= 2
h)	= 4	q)	= 8
i)	= 6		

ROD WORK AREA

Keeping On Track

i					**k**		**a**			**n**				
					p									
									f				**j**	
														l
b						**d**								
	e			**m**				**o**		**h**		**g**		**c**

Rule: White = 1
Make Problems to Match the Answers

ACROSS			DOWN		
a)	=	4	i)	=	5
b)	=	7	j)	=	3
c)	=	1	k)	=	5
d)	=	10	l)	=	2
e)	=	1	m)	=	1
f)	=	4	n)	=	5
g)	=	1	o)	=	1
h)	=	1	p)	=	4

ROD WORK AREA

My Own Puzzle

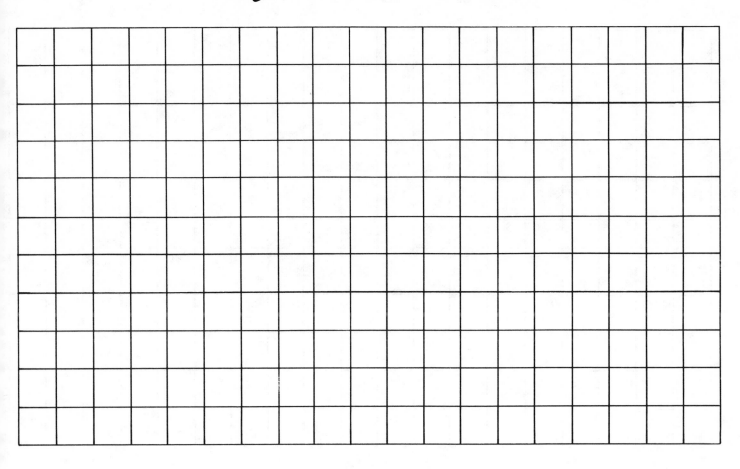

Rule: White = 1

ROD WORK AREA

Picture Puzzles with Cuisenaire Rods
© 1979 Cuisenaire Company of America, Inc.

My Own Puzzle

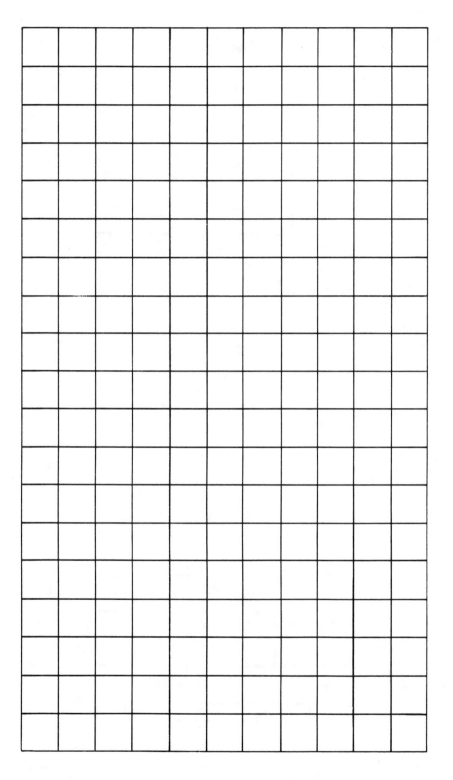

Rule: White = 1

ROD WORK AREA

ANSWERS

A Big Hello

page 6

The Open Road

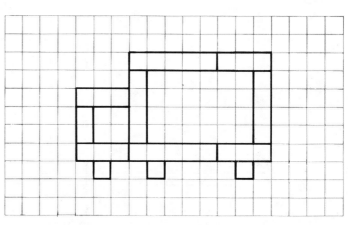

page 7

Mirror Magic

page 8

Discovering America

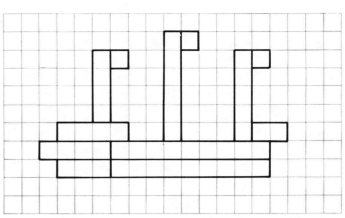

page 9

The Great Pumpkin

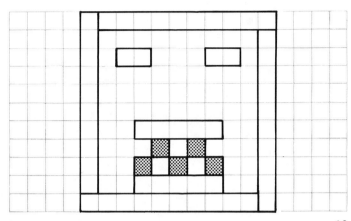

page 10

My Teacher

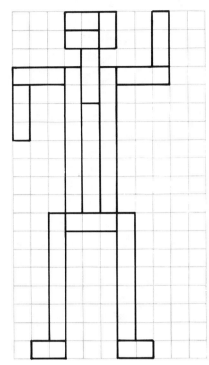

page 11

Pin-Wheel Pattern

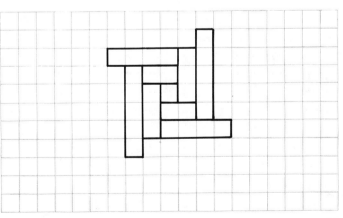

page 12

One To Eight

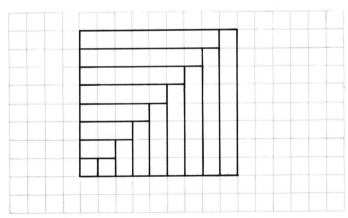

page 13

I'm Going Ape

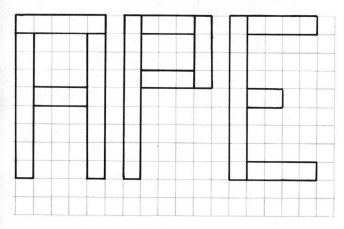

page 14

Desert Friend

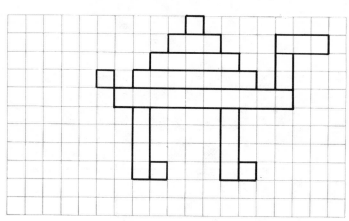

page 15

One Step At A Time

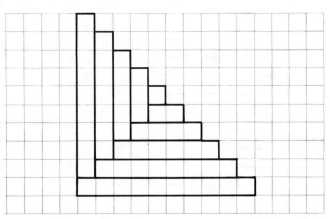

page 16

Are You My Mother?

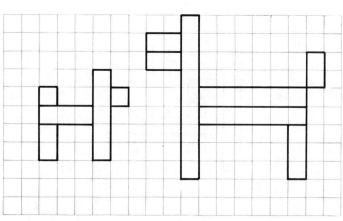

page 17

Circus Fun

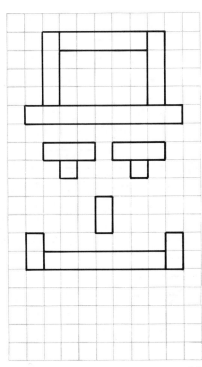

page 19

Under Water

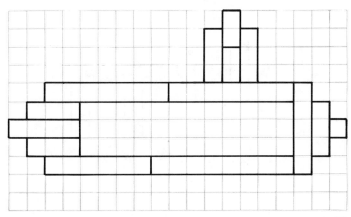

page 18

Season's Greetings

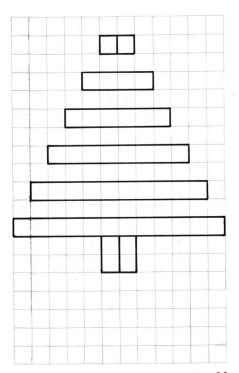

page 20

To The Rescue

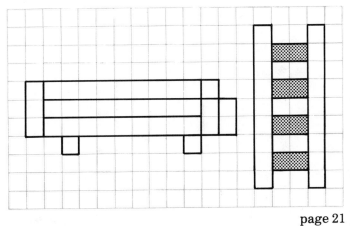

page 21

Desert Flower

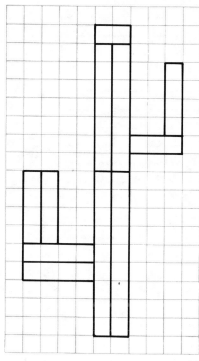

page 22

Sssss

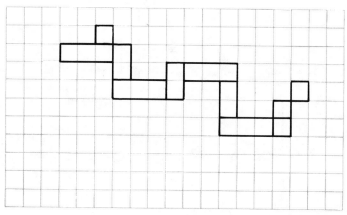

page 23

More Mirror Magic

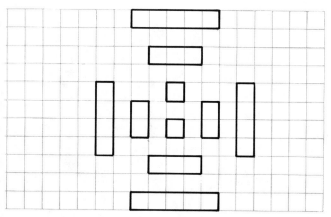

page 24

Mr. Robot

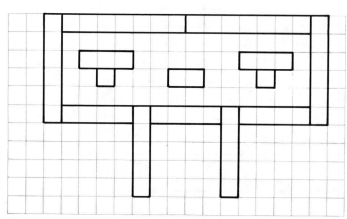

page 25

Riding In Style

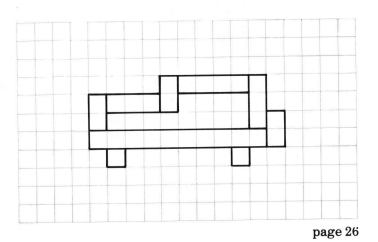

page 26

What's For Dinner?

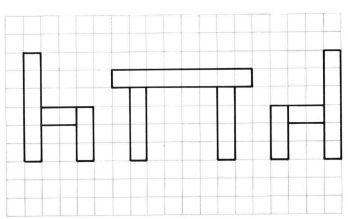

page 27

Upside Down

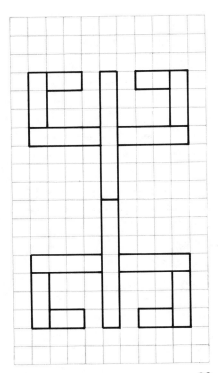

page 28

Gee Whiz

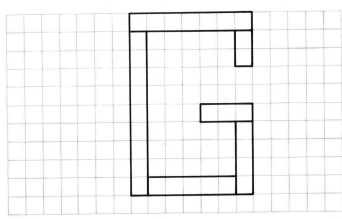

page 29

Stretching Up

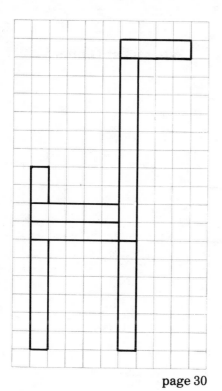

page 30

Goofy

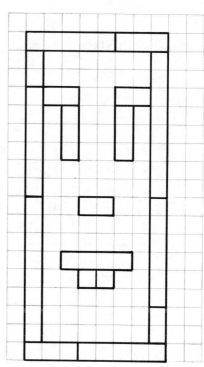

page 31

Go Fly A Kite

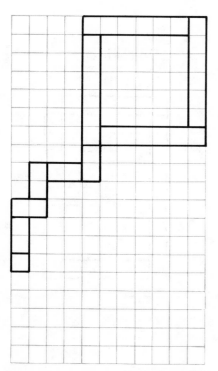

page 32

The Big Push

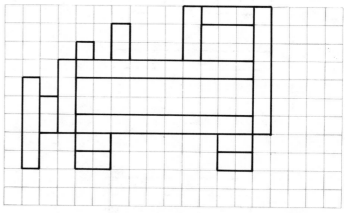

page 33

Square Deal

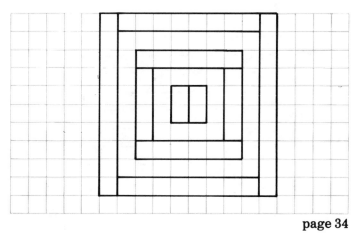

page 34

Poodle, Rhymes With Noodle

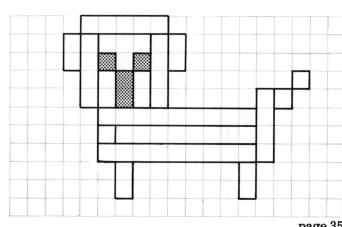

page 35

Bed Time

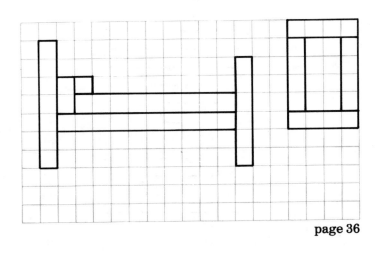

page 36

You Name It

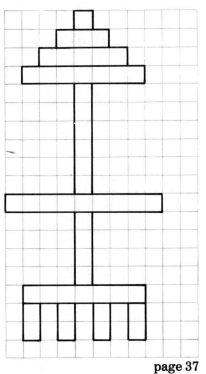

page 37

Animal World

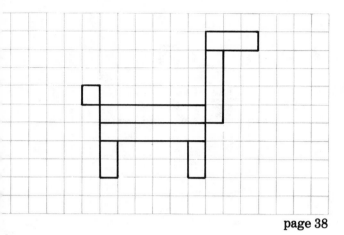

page 38

Teacher's Pet

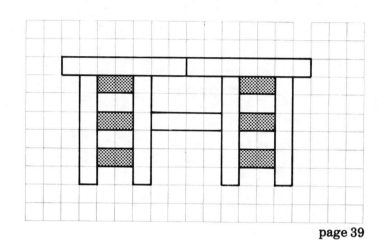

page 39

Turning Corners

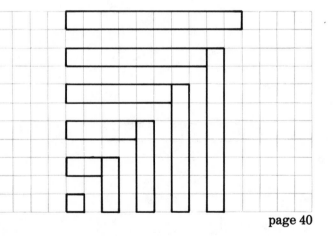

page 40

A-Maz-ing

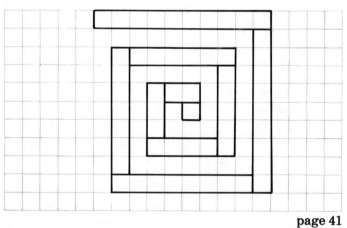

page 41

April Fools (Only If You Do Page 41)

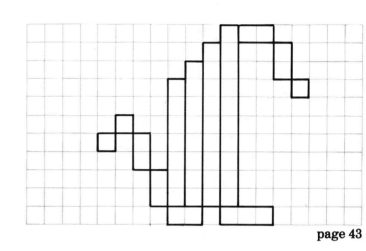

page 42

Puff, The Magic Dragon

page 43

To And Fro

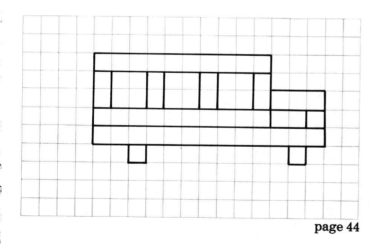

page 44

Be My Valentine

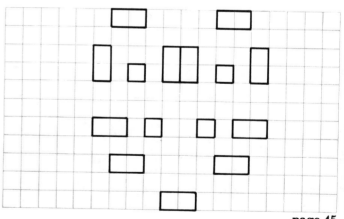

page 45

60

Once A Year

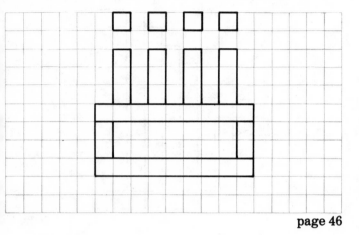

page 46

Quilt Pattern

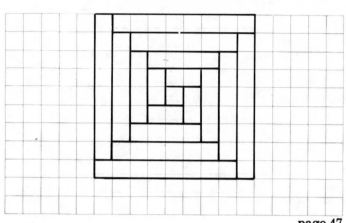

page 47

Keeping On Track

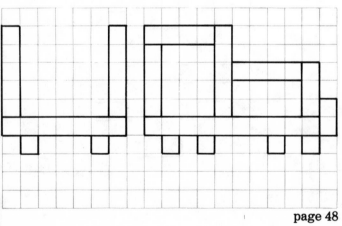

page 48